SCIENCE ENCOUNTERS

Music

JULIAN ROWE

Published by Rigby Interactive Library, an imprint of Rigby Education,
division of Reed Elsevier, Inc.
500 Coventry Lane
Crystal Lake, IL 60014

Interiors designed by **AMR**
Illustrations by Art Construction

Printed in the United Kingdom

Library of Congress Cataloging-in-Publication Data
Rowe, Julian
Music / Julian Rowe.
p. cm. -(Science encounters)
Includes bibliographical references and indes.
Summary: Describes various musical instruments and uses familiar examples to explain the scientific principles of music.
ISBN 1-57572-091-4
1. Music-Acoustics and physics-Juvenile literature. 2. Musical instruments–Juvenile literature. 3. Sound–Juvenile literature.
[1. Music. 2. Musical instruments.] I. Title. II. Series
ML3928.R69 1997
781.20–dc20
96-17979
CIP
AC MN

Acknowledgments
The publisher would like to thank the following for permission to reproduce photographs.

Rex Features London, p. 5 (top); Clive Barda/P.A.L., p. 5 (bottom), p. 12, p. 16, p. 18, p. 19; Tony Stone Images/J. Sneezby/B Wilkins, p. 7; Tony Stone Images, p. 8, p. 11, p. 14; Jonathan Fisher/P.A.L., p. 13; Colin Willoughby/P.A.L., p. 15, p. 17; The Bridgeman Art Library, p. 20; Hulton-Deutsch Collection, p. 22; Barnaby's Picture Library, p. 23, p. 24; Rachel Hughes/P.A.L., p. 26; Pictorial Press Limited, p. 28

Every effort has been made to contact copyright holders of any material reproduced in this book. Any omissions will be rectified in subsequent printings if notice is given to the publisher.

CONTENTS

SCIENCE IN MUSIC

The world is a noisy place. We hear all kinds of sounds, from traffic noise to the softest music. We hear people talking. We hear sounds that have special meanings—a telephone ring, a fire alarm, a doorbell. All sound travels through the air as sound waves. Everything we hear is the result of these sound waves entering our ears. But just what are sound waves?

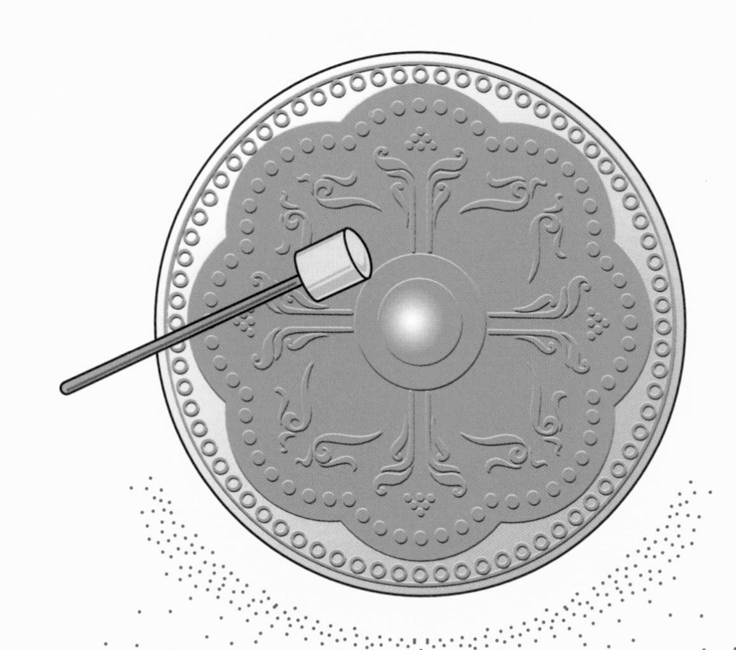

Waves of Sound

When a gong is struck with a hammer, the gong vibrates. While it continues to vibrate, we hear the musical sound the gong makes. The gong's vibration forces the air **molecules** around it to vibrate backward and forward. This causes air molecules around the gong either to crowd together or to move apart rapidly. Where they crowd together they create areas of high **air pressure.** Where they move apart there are areas of low air pressure. These areas spread outward from the source of the sound—the gong. They are the sound waves that the ear hears.

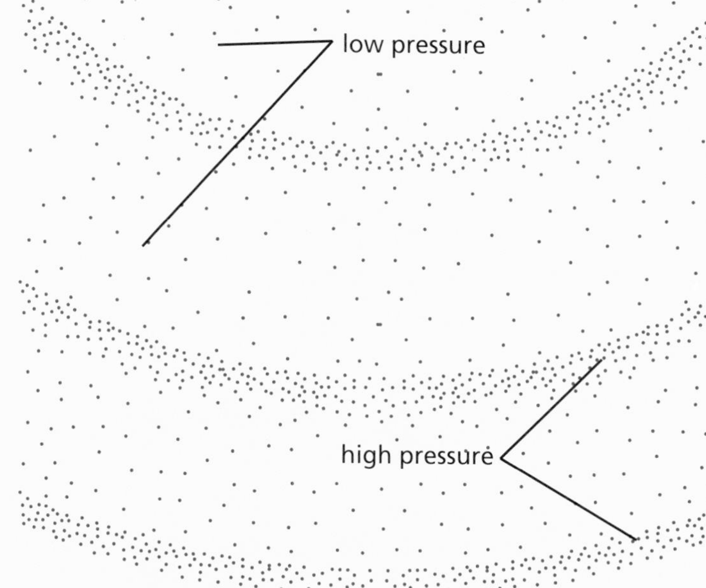

The distance between one area of high air pressure to the next area of high air pressure is equal to the wavelength of the sound. A low sound has a longer wavelength than a high sound; a lower sound also has a lower **frequency**— there are fewer complete waves of sound each second.

All Kinds of Instruments

Music is the arrangement of sounds into patterns that are interesting or pleasing to hear. If the sounds are not pleasant, the result is noise! People play music for many reasons —for relaxation, for entertainment, and to express their feelings. Music—pop, classical, or religious—is part of everyone's culture. Music can be as simple as the beating of a rhythm on a drum or as complicated as a musical performance involving hundreds of singers and a full orchestra of stringed, wind, percussion, and keyboard instruments.

The clear ringing sound of a steel band is typical of Caribbean music. The tuned drums are made out of metal oil barrels, cut in half. Shallow, circular dents are beaten into the flat top or bottom sections of the drums. Each dent causes a different sound. The larger drums play the lowest sounds of all.

With the invention of radio, records, and tapes, more people now listen to more music than at any time in history. New sounds and ways of making music—using **electronic** instruments and computer techniques—are constantly being explored.

When a singer sings a high note she can be heard at the back of a large auditorium. The musical sound that a human voice makes is produced by vibrations of parts of the throat called the vocal cords.

YOUR EARS

The sounds you hear all around you travel by way of air particles into your ears. High sounds, like a whistle or a scream, are caused by air particles that vibrate very quickly. Low sounds, like a distant rumble of thunder or a growl, are caused by slow vibrations in the air. Sound is a form of energy, and your ears are marvellous devices that change sound energy into sounds that you can hear. Have you ever wondered how you hear sound?

How You Hear

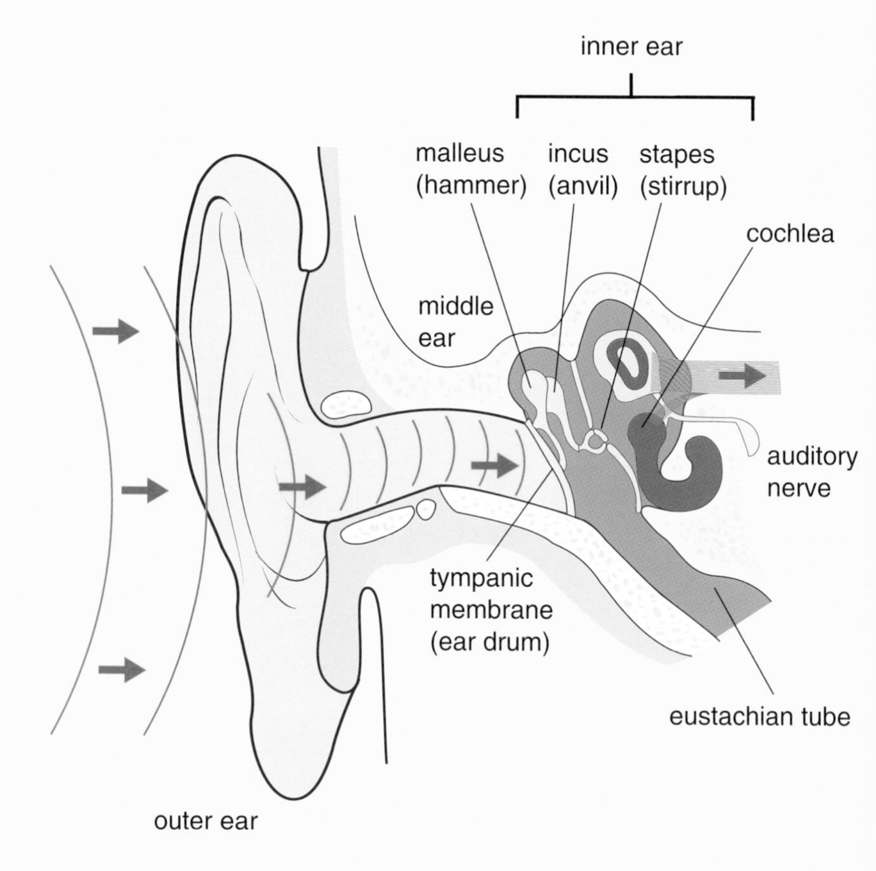

Your ear has three main sections. The outer ear, or auricle, is the part of your ear that you can see. It collects sounds and moves them down a short tube inside your ear, called the auditory, or ear, canal.

At the end of your ear canal, about 1 inch inside your ear, a membrane (thin sheet of tissue) called the eardrum stretches over the opening to your middle ear. Sound waves hit your eardrum and make it vibrate.

These vibrations are picked up by three tiny connected ear bones called ossicles. (One of them, the stirrup, is the smallest bone in your body.) The ossicles pass these vibrations to the cochlea, a fluid-filled spiral tube in your inner ear. Inside your inner ear, movements of the fluid cause tiny, sensitive hairs to vibrate. Their movement makes electric signals, which the auditory nerve sends to your brain. Your brain decodes these signals, and you hear a sound.

Sound, or air vibrations, are converted to electric signals in your ear. The same thing happens inside a **microphone!**

What You Hear

The frequency of sound waves is the number of vibrations that a sound makes every second. Scientists use a unit called the **hertz** (Hz) to measure frequency. One hertz (1 Hz) is one complete wave every second. Human beings can hear sounds between 20 Hz (a low rumble) and 20,000 Hz (some whistles). Many adults lose their ability to hear high notes as they get older. Dolphins, however, can hear sounds with a frequency of an astonishing 150,000 Hz. Whales use low-frequency sounds to communicate with each other across long distances in the oceans.

A dog can hear far higher sounds than we can. In fact, there are special dog whistles that only dogs can hear.

SOUND AND RADIO WAVES

Two German scientists made important discoveries in sound. Herman Helmholtz (1821–1894) studied sight and hearing and showed how the cochlea works. His student, Rudolf Hertz (1857–1894), discovered electromagnetic (radio) waves.

SOUND BEHAVIOR

Do you like singing in the bathtub? The sounds you make echo, or reflect, off the flat bathroom tiles, and there is nothing soft nearby to absorb the sound. Nearly anyone can sound like an opera star in a bathroom! But echoes, and materials that absorb sound, can be a problem in a large auditorium. They cause some sounds to fade away altogether. It is difficult to design a concert hall so that all instruments and speech can be clearly heard by everyone in the audience. How have scientists solved this problem?

SPEED OF SOUND

Sound travels through air at a speed of 1,116 feet per second. Fighter planes regularly fly faster than the **speed of sound** through the air (about 760 miles per hour).

The Whispering Gallery in the dome of St. Paul's Cathedral in London is famous because of its **acoustics.** A whisper near the wall on one side can be clearly heard on the other side, 107 feet away. Ordinarily this would be impossible, but the circular shape reflects the smallest sound.

Listening to Music

In a Gothic church that has a high vaulted ceiling, even quiet sounds such as a whisper echo and reverberate (echo over and over again). These effects make music come alive. But if there are too many echoes you can't hear anything clearly. If there are none at all the music sounds dead. When performers rehearse in empty theaters or concert halls they can hear the difference. When there is an audience present, clothes and bodies absorb sound, cutting down the reflection of sound.

Making Sounds

A musical instrument makes a sound, called a tone, when part of the instrument vibrates. It could be a string, the skin of a drum, the wooden bar of a xylophone, or a moving column of air inside a wind instrument. These vibrations make sound waves. The instrument contains the sound waves so that the player can control them and make them sound a certain way.

Sound waves can be made visible using an instrument called an oscilloscope. It shows pictures of sound waves on a small TV screen. A very pure note, like that made by a tuning fork, produces a very smooth, regular wave pattern. A gong makes a jagged, irregular pattern, and a violin note makes a complicated but regular pattern.

Sound Reflections

Echoes are reflections of sound waves. They are like a mirror for sound. When you shout loudly, the sound you make travels outward from you. If it meets a hard, flat surface, such as the wall of a room, the sound is reflected—it changes direction. If you are outside near a building, a wall, or a mountain, you may also hear echoes.

VOICE TRICKS

If people swallow a mouthful of helium from a party balloon and then talk, they sound very silly! This is because sound travels at different speeds through different substances—gases, liquids, and solids—and so changes.

In the mountains, you can hear your voice echo back to you when you shout. The sound bounces off a surface such as a cliff, and you may hear several echoes before the sound dies away. Because sound travels at the same speed all the time, the time an echo takes to return tells you how far away the cliff is.

MUSICAL NOTES

Have you ever heard some music and then said afterward, "I can't get that tune out of my head"? You can remember the melody (tune) and its rhythm. A composer also uses harmony and tone color and chooses the notes to make a good piece of music. But how do composers pass onto musicians the sounds or music they have in their heads?

Writing It Down

Most music is based on a scale, or a set of tones arranged by **pitch.** The Ancient Greeks were the first people to make letters stand for musical tones, or notes, of a scale. The system of writing down notes (notation) has changed little since the 1400s. Lined music paper was invented much later. In many countries notes are named after the first seven letters of the alphabet—A, B, C, D, E, F, and G.

The five lines on a musical score are called a stave. It is divided by vertical (up and down) lines into bars (groups of beats) where the musical notes and rests are written. The rests show pauses (how long not to play) and the different forms of notes show how long the notes should be played. Where each note appears on the stave tells which note should be played.

PYTHAGORAS' PITCH

People first tried to understand music using science a long time ago. For example, Pythagoras, the famous Greek mathematician (c. 582–507 BC), investigated the pitch of a note created by a stretched string. The pitch of a note describes how high or low it is. He discovered that pitch depends exactly on the length of the string; the shorter the string, the higher the pitch.

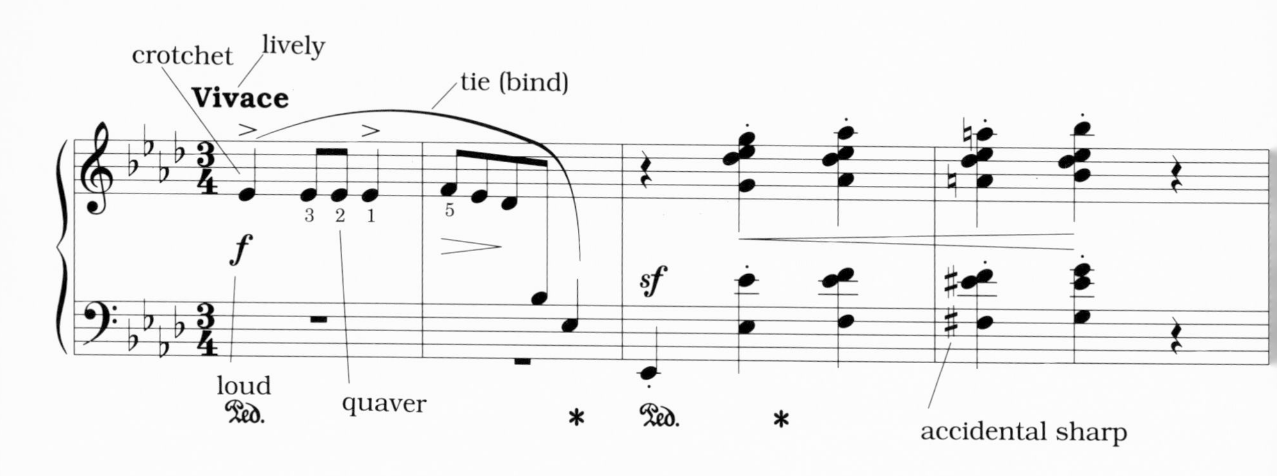

One piece of music can be divided into different parts for different instruments. This way musicians can all play together in harmony.

Tone Color

A piano sounds quite different from a trumpet. Why? Musical notes have a special quality that helps a listener to tell which instrument is playing. It is called timbre or tone color. The sound of each instrument has a different timbre. Although the main frequency of the same note on different instruments is the same, each instrument colors every note with characteristic sound.

MEASURING SOUND

Scientists use a unit called a decibel (dB) to measure the "loudness" of a sound. The smallest, or least intense, sound that a normal human ear can hear is 0 decibels. In contrast, a jet plane taking off creates about 120–130 dB, the same as music that is amplified at a rock concert. The sound of rustling leaves is about 30 dB.

WHISTLES, PIPES, AND FLUTES

Have you ever made a musical note by blowing across the neck of a bottle? Like a bottle, most wind instruments consist of a hollow tube. But they also have a mouthpiece. Wind instruments have been around since ancient times. Whistles carved from the toe bones of reindeer 40,000 years ago have been found in France. They make sounds when the air inside the hollow tube vibrates. There are two kinds of wind instruments: woodwind instruments and brass instruments.

When you blow into or across the mouthpiece of a woodwind instrument, such as a flute or a whistle, you make the air inside the instrument vibrate. This movement of air produces the note. By covering the holes in the instrument with your fingers, you can change the length of the column of air that vibrates inside the instrument. This changes the pitch of the note.

Panpipes produce a haunting, breathy sound and are associated with the mythical Greek god Pan. Deeper notes are made by blowing across the longer pipes; higher pitched notes are produced by blowing across the shorter pipes.

A FAMILY OF INSTRUMENTS

Most woodwind instruments were at one time made of wood. Today many are made of metal. The concert flute makes a bright, clear sound. Like the flute, the piccolo is held sideways but makes a higher sound. The bright notes of the fife, a small side-blown flute, sound high above the rhythms of a marching band.

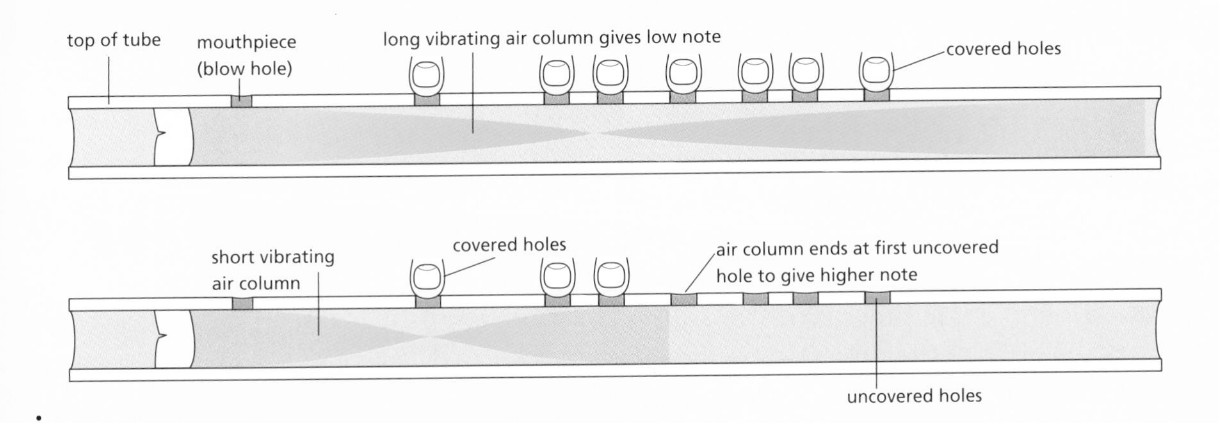

When a flautist (flute player) covers all the holes of the flute, the note produced is the lowest possible, because the column of vibrating air is long. When some of the holes are uncovered, a higher note is sounded. The column of vibrating air is now shorter. The same principle is used to change the pitch of the sound in other wind instruments, such as the oboe, clarinet, and bassoon.

FLAUTISTS' FINGERS

Theobald Boehm (1794–1881), an inventor and flautist, set out to make the perfect flute. He soon found that he would have to make holes in places on the flute where players' fingers could not reach. So he made the key mechanism that modern flutes still use today.

The complex keywork of a modern metal concert flute makes it easier to play than its simple wooden ancestors. And today's flute has a much brighter sound. Some of the pads are closed directly by the flautist's fingers and others by keys operated by the thumbs or little fingers.

LISTENING TO THE BAND

Brass instruments, the other wind instruments, make a fantastic sound. Most have a mouthpiece at one end of a hollow tube and a flared bell at the other. The sound made by the player's lips vibrating against the mouthpiece is the **fundamental** note of the instrument. By changing the tightness of the lips and by blowing harder or softer, the player raises or lowers the pitch of the note. This is the only way to play different notes on some instruments, such as a hunting or coach horn.

Blazing Fanfares

A fanfare on a trumpet is a blaze of sound that can be clearly heard above any other instrument. The trumpet's hollow tube is coiled. A system of valves directs air through a shorter or longer length of the instrument.

Sometimes a trumpet player puts a round object called a mute inside the bell of the instrument. A mute does not silence the trumpet. It makes it quieter and gives the sound a special quality. Mutes with different shapes cause different sounds. They can produce a jazzy, buzzy quality or produce a thin piercing wail.

All the different brass instruments are played in a military band, as well as some woodwind instruments and a variety of percussion instruments.

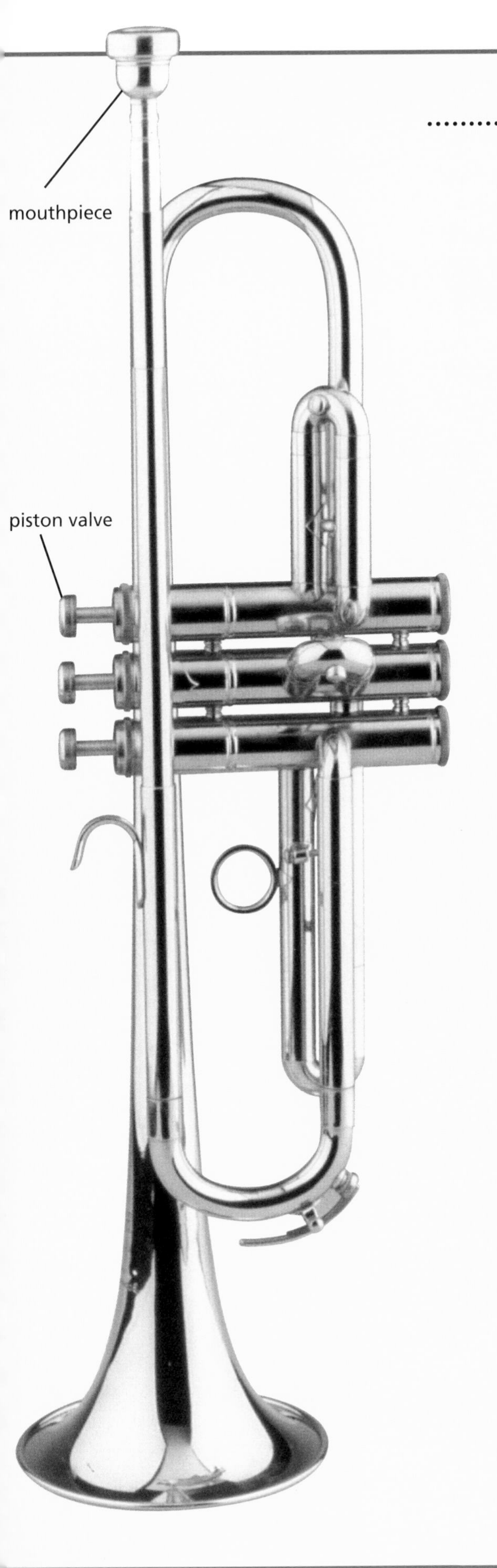

A trumpet has three valves that open up or shut off passages for the vibrating air. As for all wind instruments, the longer the passage or tube, the lower the note.

THE SAX

The saxophone was patented in 1834. It was used first in military bands before it became a leading jazz instrument. Although it is a wind instrument, it has a very different mouthpiece. Instead of a fixed brass "tube," to blow into, the saxophone has a reed—sliver of bamboo—that the player puts inside the mouth. It is the vibrating reed that causes a column of air inside the instrument to vibrate.

PHARAOH'S TRUMPETS

Excavations in 1923 of the tomb of the Egyptian pharaoh Tutankhamen uncovered trumpets buried with him. The instruments that were found still could be played.

THE OLDEST INSTRUMENTS

When you think of percussion instruments, do you hear the crash of cymbals and the rhythm of drums? There are many different kinds of percussion instruments. In fact anything that makes a sound when you hit it or shake it can be used as a percussion instrument. Maracas, for example, are made of beads or lead shot inside wooden shells or hollow gourds. They are played by shaking them to produce a repetitive rhythm for a musical group or singer. Percussion instruments are used in music all over the world.

Playing the Bars

Each wooden bar of a xylophone makes a different note. The bars are arranged in the same way as the black and white keys of a piano. The pitch of each note varies with the length of the bar. Shorter bars produce higher notes. The notes are amplified by hollow metal tubes that hang underneath, which resonate. **Resonance** happens when one vibrating object, in this case the xylophone bar, causes something else, the tubes, to vibrate.

A vibraphone, like a xylophone, is played by striking the bars with a beater. The bars are metal and the beater has a soft head. Metal tubes that contain electric fans are positioned under the instrument and produce a **vibrato** effect.

PERCUSSION IN TUNE?

Xylophones, tubular bells, and kettle drums are examples of instruments that can be tuned, or adjusted for pitch. African "talking drums," can be tuned to produce high and low notes. Some instruments, such as cymbals, castanets, triangles, and some drums, are not tuned to play a particular note.

Evelyn Glennie is the first full-time solo percussionist in the world. A remarkable musician who is also hearing impaired, she owns more than 700 percussion instruments, which she collected from all over the world. Here she is performing on the xylophone.

Beating the Drum

Drums are played all over the world, and they all have the same basic design. The drumhead, the drum's most important part, is made of animal skin or plastic stretched over the hollow body of the instrument. Some drums are bowl-shaped, like the tympani of an orchestra; others are tube-shaped. Tambourines are a kind of drum with a round open frame. Drums are tuned by adjusting the **tension** of the drumhead. Tension screws or levers at the side of the drum can be tightened to stretch the drumhead more. This raises the drum's pitch. To lower the pitch, the screws or levers are loosened.

A basic drum kit consists of a bass drum, tom-toms, a floor drum (tenor drum), a snare drum, and cymbals. Drummers use different beaters—hard-headed and soft-headed sticks and a wire brush. The floor drum is operated by pedals.

KEYBOARDS

A keyboard is like a set of levers that link each movement of the player to the production of a note. A piano player makes a different note with each finger and can control the loudness of the sounds with foot pedals. An electronic keyboard uses electronics to copy the sounds of many different instruments, including a piano.

An organ is a keyboard instrument with pipes. Instead of hammers hitting strings, air goes into the pipes and vibrates in them, like air in a flute.

The Piano

If you were to look inside a piano you would see a long line of metal strings. The strings are shorter at one side and longer toward the other side. High notes are made by shorter strings. Low notes are made by longer strings. Some of the strings are much thicker and heavier than the others. They, too, make lower notes. A piano tuner tightens or loosens the piano strings until they have exactly the correct pitch.

The sound of a piano comes from a wooden soundboard located under the strings. A vibrating string by itself does not make a loud sound. The vibrating wood of the soundboard helps to amplify the sound of the strings.

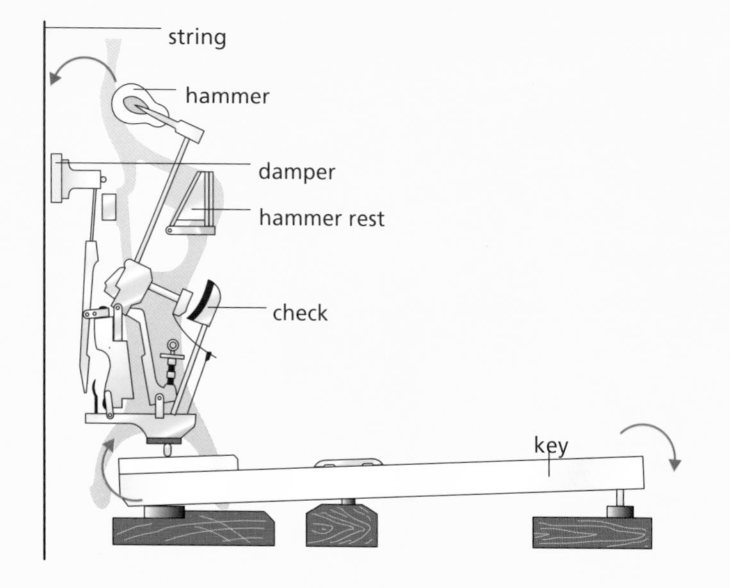

Piano keys move felt-covered wooden hammers. These strike the strings. When the keys are at rest, dampers press on the strings to keep them from vibrating. The mechanism that links the key to the hammer also raises the damper as the key is played.

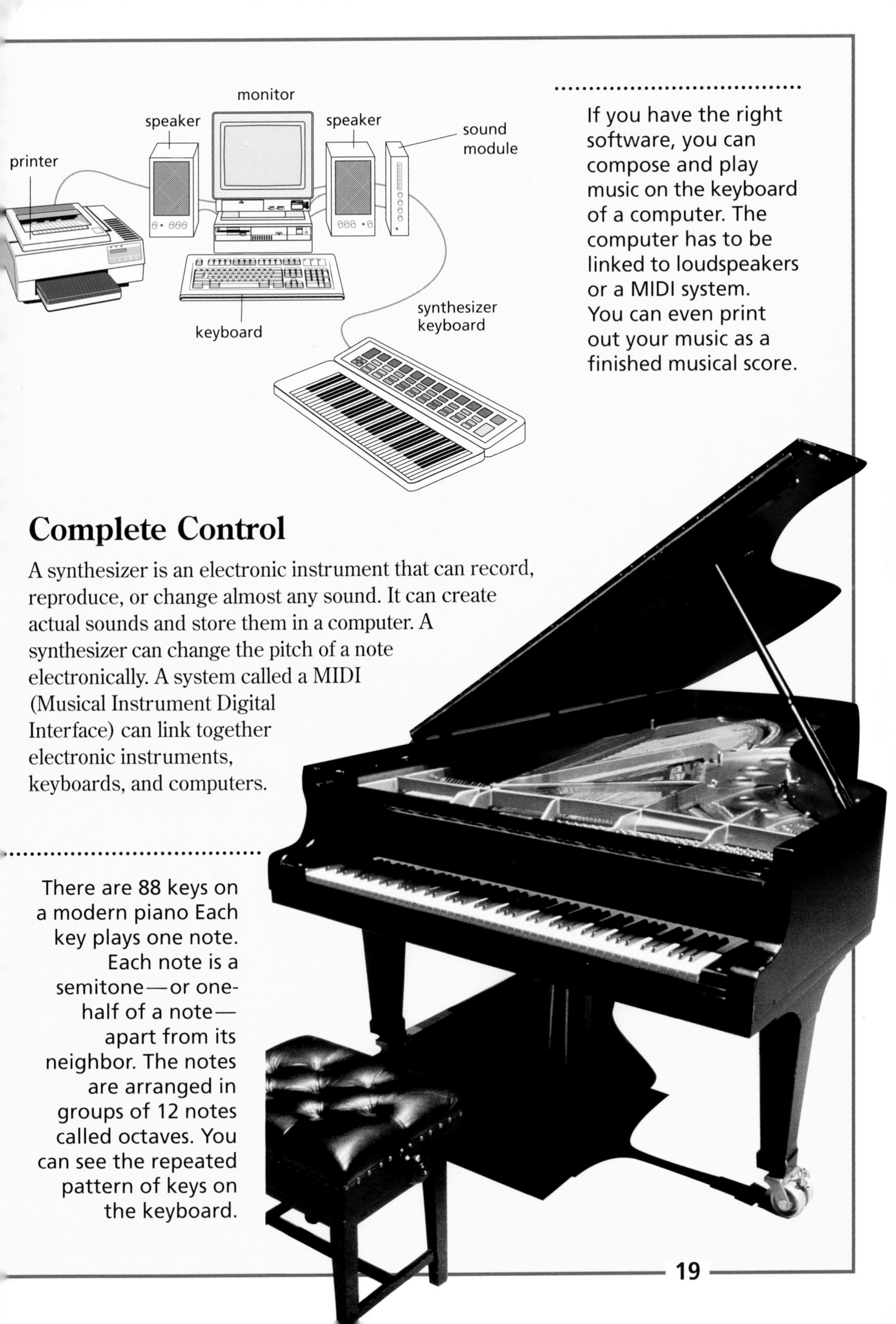

If you have the right software, you can compose and play music on the keyboard of a computer. The computer has to be linked to loudspeakers or a MIDI system. You can even print out your music as a finished musical score.

Complete Control

A synthesizer is an electronic instrument that can record, reproduce, or change almost any sound. It can create actual sounds and store them in a computer. A synthesizer can change the pitch of a note electronically. A system called a MIDI (Musical Instrument Digital Interface) can link together electronic instruments, keyboards, and computers.

There are 88 keys on a modern piano Each key plays one note. Each note is a semitone—or one-half of a note—apart from its neighbor. The notes are arranged in groups of 12 notes called octaves. You can see the repeated pattern of keys on the keyboard.

ONE, FOUR, OR MORE STRINGS

When an archer shoots an arrow, the string of the bow makes a musical twang. At first, lyres were little more than "musical bows." Harps have been found by archaeologists looking for traces of the earliest civilizations.

There are two main types of stringed instruments: those like the violin, which are played by drawing a bow across the strings, and those like the harp or guitar, which are played by plucking the strings.

Notes and Harmony

Players of a stringed instrument such as a violin, sitar, or a guitar produce different notes by pressing on the strings with their fingers. This changes the strings' lengths. If a string is divided in two, the note will vibrate twice as quickly—it has twice the frequency of the original note. Every note has a different frequency.

LUTES

The lute is an ancestor of the guitar and violin. Originally from Arabia, it has a pear-shaped body and 11 strings. The player plucks the strings. There are many different kinds of lutes —the South American charango, the Japanese shamisen, the Greek bazouki, the Russian balalaika, and the Indian sitar are some of them.

This 15th century manuscript painting shows that lutes were played many centuries ago.

Sounding Good

The guitar was already popular by the 17th century. There are two kinds of modern guitar: the acoustic (or Spanish) guitar and the electric (or rock) guitar. To tune a guitar, the tuning heads are turned to change the tension of the strings. The upper part of the body of a guitar is the soundboard. It is made from two pieces of wood glued together. The soundboard gives the six-stringed acoustic guitar its special sound.

tuning head

The electric guitar usually has a solid body and devices called pickups that are mounted under the strings. These convert the vibrations of the strings into electric signals. These signals are amplified and heard through a loudspeaker. The strings that play the lowest notes are the thickest.

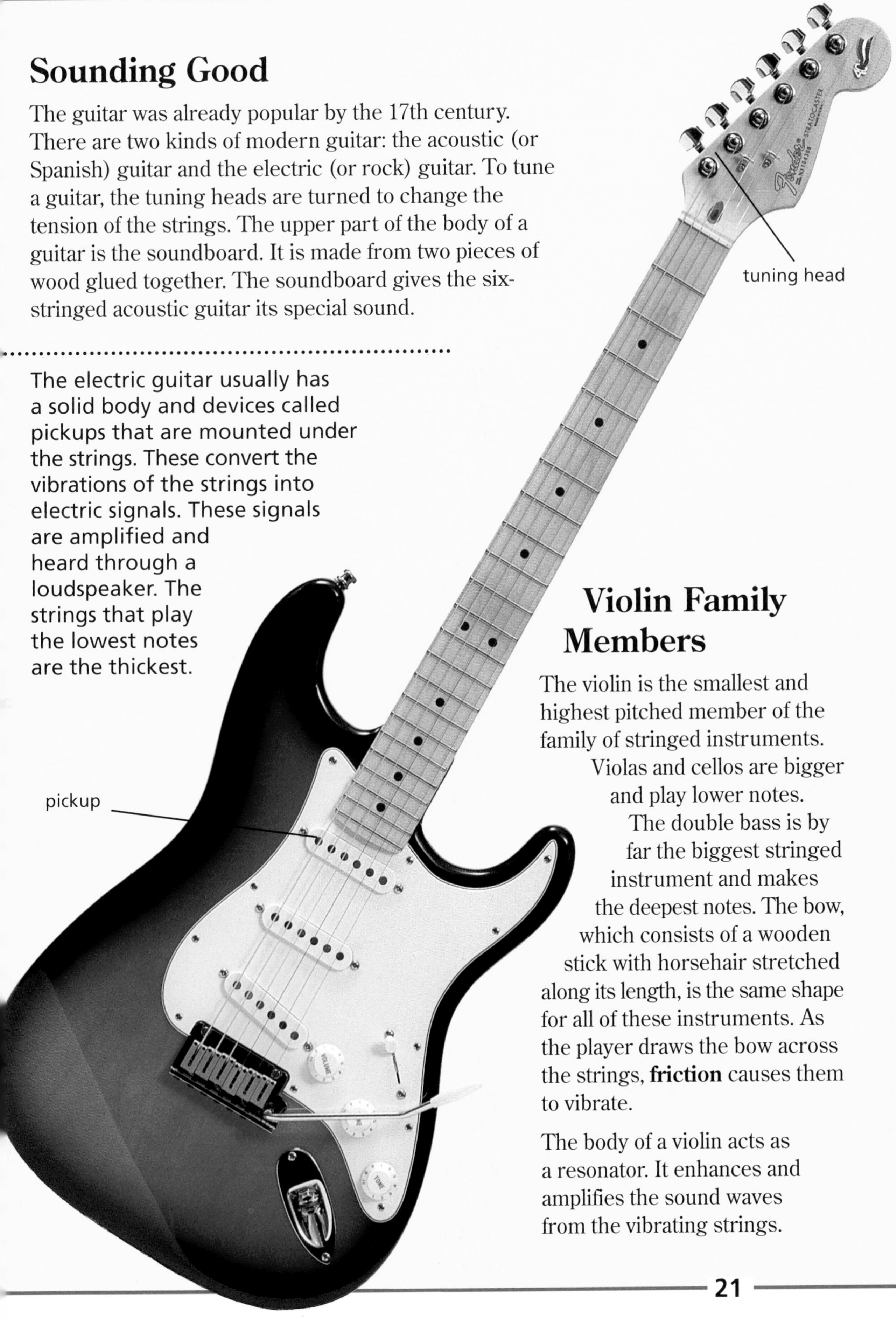

pickup

Violin Family Members

The violin is the smallest and highest pitched member of the family of stringed instruments. Violas and cellos are bigger and play lower notes. The double bass is by far the biggest stringed instrument and makes the deepest notes. The bow, which consists of a wooden stick with horsehair stretched along its length, is the same shape for all of these instruments. As the player draws the bow across the strings, **friction** causes them to vibrate.

The body of a violin acts as a resonator. It enhances and amplifies the sound waves from the vibrating strings.

RECORDING MUSIC

Your two ears tell you where a sound comes from. They give it "depth." When you hear recorded music coming from only one loudspeaker, it seems flat and uninteresting. The music is **monaural,** which is like listening with only one ear. With a **stereophonic** recording the music sounds as though it is live. Two or more microphones are needed to record stereophonic music. How are music recordings made?

Making a Record

Recording involves making a permanent copy of sound waves so that they can be heard again. On a compact disc (CD), this copy is stored in tiny pits (dents) in the disc. On tape, the sounds are copied onto bands, or tracks, of magnetism. On **vinyl** records, the sounds are stored in the spiral groove on the surface of the record. If you look at the groove with a magnifying glass you can see a pattern of waves. Deep waves produce low sounds; lots of waves mean high sounds.

The First Record

Thomas Alva Edison (1847–1931) invented the phonograph — the first record player—in 1877. To record sounds, one turned a handle to rotate the cylinder covered in tin foil and spoke into the mouthpiece. Inside the mouthpiece was a thin disc of metal, with a metal needle attached to its center. As sounds made the metal disc vibrate, the needle scratched a pattern on the foil. To listen to the recording, you reversed the process. As you turned the handle, the scratches in the tin foil made the needle vibrate the metal disc, reproducing the sounds.

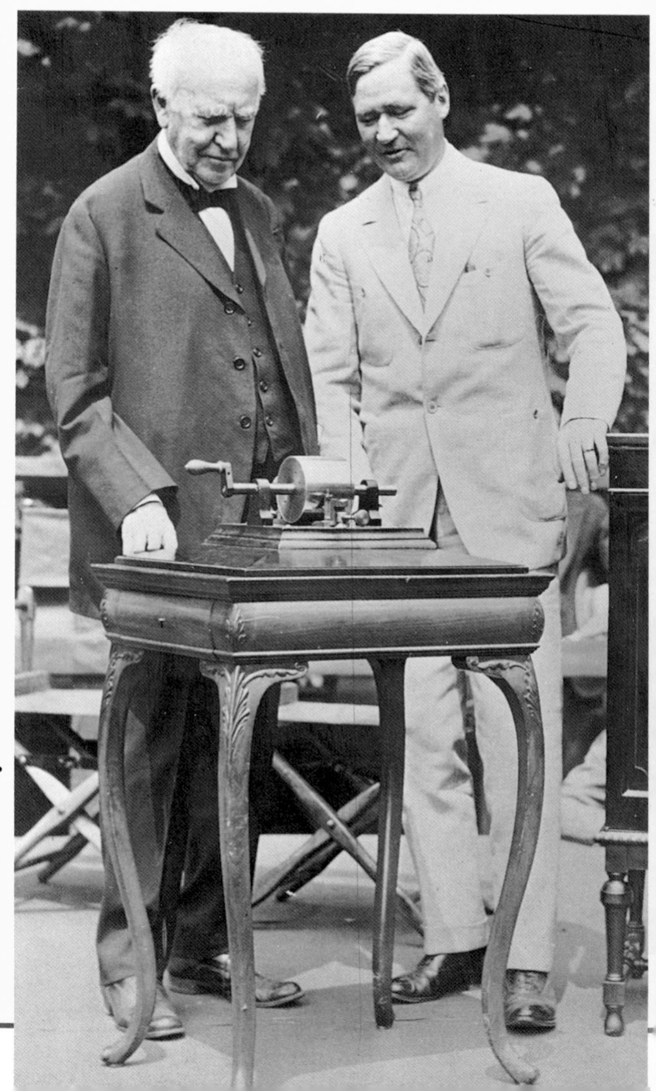

The first words recorded by Edison on his phonograph were "halloo, halloo." The world's first talking machine sold for $18.

Electric Records

After 1925 records could be played electrically. The needle, or stylus, which cut the master record could now be controlled by an electric current.

When a record is played the stylus traces the pattern of waves in the groove as the record rotates. The movement of the stylus generates electric signals. These are amplified electronically and fed into a loudspeaker. The loudspeaker changes the signals back into sounds.

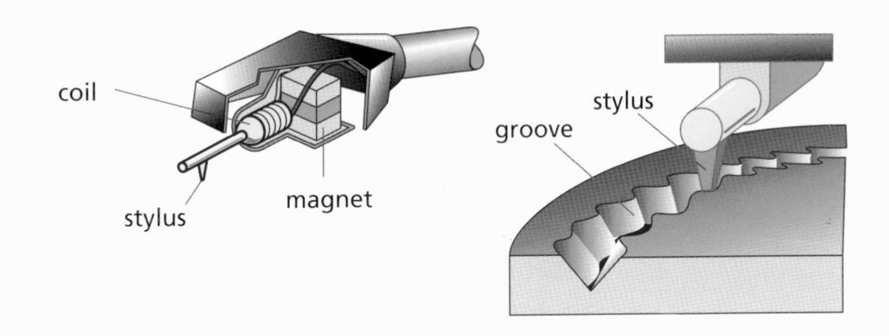

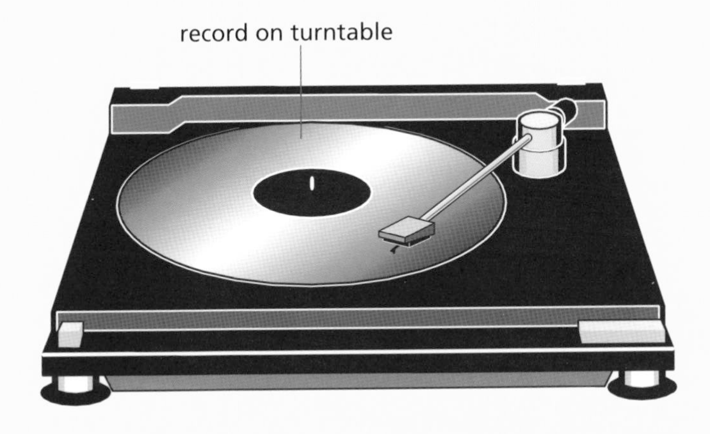

On a stereo record the stylus traces two sets of sound patterns, or tracks. These are positioned on opposite walls of the groove.

In a recording studio, the different instruments are recorded separately. The sounds are captured by 24 or more microphones on separate tracks. The sound tracks are then "mixed" to obtain a good balance. For a stereophonic recording, two tracks are recorded onto one master tape, from which all subsequent tapes, records, or CDs are made.

MAGNETIC TAPES

To make a CD, audiotape, or record, a producer first has to record the music onto a tape. The whole process depends on magnetism and electricity. When an electric current flows in a wire, it acts like a magnet. If you place a compass near the wire, you will see that the compass needle no longer points in a north–south direction. If you wind the wire around a piece of iron, the iron will become a strong magnet as long as the electric current flows. If you move a magnet near a coil of wire, an electric current will flow in the wire. This is **electromagnetism** at work. But how does it help in tape recording work?

The Microphone

During a recording, the sound is first captured by a microphone. One kind of microphone is the moving-coil microphone. It consists of a thin, round plate of metal attached to a coil of wire. The coil is close to, but does not touch, a small magnet. When a sound makes the metal plate vibrate, the coil of wire moves. The magnet causes an electric current to flow in the coil. The current is amplified and used to make the recording on tape.

A battery-powered personal stereo offers high-quality sound while you are on the move. Earphones deliver sounds from the tape to your ears.

MAKING MAGNETS

You can magnetize a needle by stroking it with a magnet. The magnetic field (which is very strong) creates a magnetic effect in the needle. Before the magnetic effect of an electric current was discovered, all magnets were made in this way.

Recording on Tape

Plastic recording tape is coated with a thin layer of magnetic grains. The grains are made of iron or chromium metal. On a blank tape, the magnetism of these grains is completely mixed up. When a recording is made, an **electromagnet** arranges the magnetic grains into a neat pattern. The result is a copy of the original sounds. They appear on the tape as bands of strongly magnetized and less magnetized tape.

Playing It Back

On a stereophonic tape there are two separate magnetic channels, or tracks. In order to hear both tracks, the tape has to run past two coils of wire. Each coil is wrapped around a separate iron core. The tiny bands of magnetized tape cause minute electric currents to flow in the coils of wire. These electrical signals are amplified and fed to separate loudspeakers or earphones, which change the signals back into the sounds that you hear.

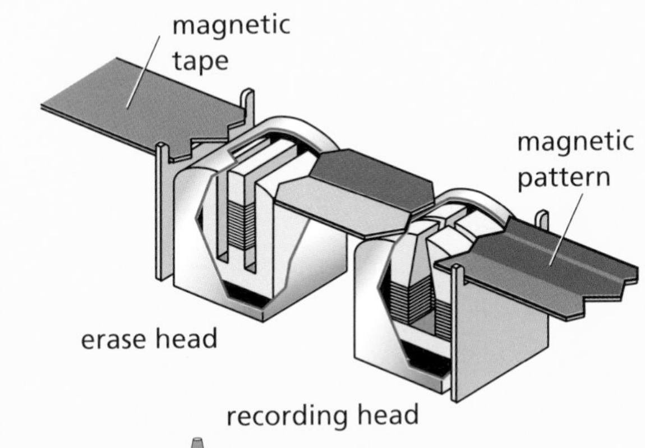

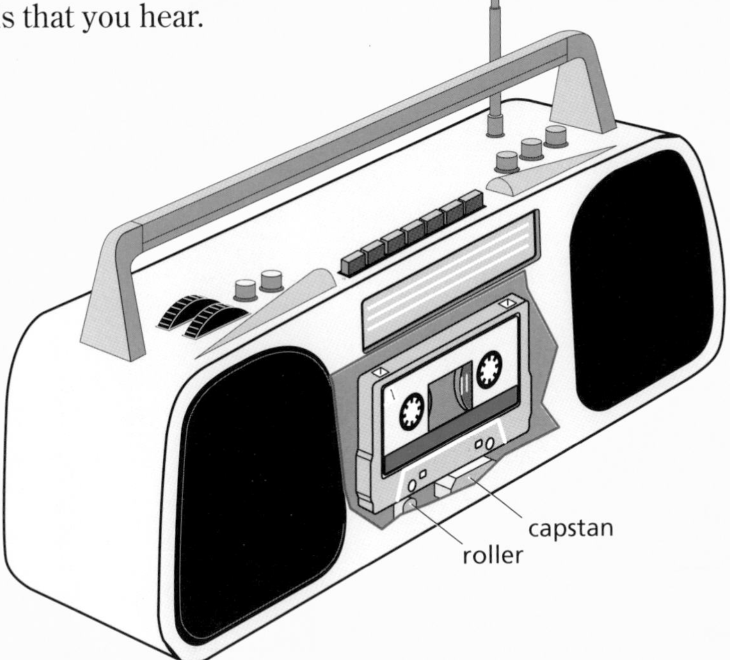

When the cassette is put into a tape recorder, a motor-driven capstan and a roller keep the tape moving at the right speed over the tape head—an electromagnet which both plays back and records on the tape.

COMPACT DISCS

The music on a compact disc, or CD, is recorded by a **laser.** Another tiny laser in the CD player plays back the recorded sounds. How does all this work?

One Step at a Time

When a CD is made, sound waves are translated into electric signals. A device called an encoder divides these signals into 44,100 parts per second of sound. Each part is given a precise number, or **digital** code, so it can be accurately recorded. This is a digital recording. A digital sound recording can be played back so that it sounds exactly like the original.

A laser uses the digital code to make microscopic pits and flat areas on the surface of the CD, which correspond to the sound waves. A CD has at least 3 billion tiny pits.

A master disc made in this way is used to make a stamp, from which all the other discs are manufactured.

On each 4-3/4-inch CD there is a spiral track three miles long that can record more than one hour of music.

Light Music

Inside a CD player, a laser reads the digital code at a rate of about 20,000 signals per second. A beam of light from the tiny laser in the CD player focuses onto the surface of the disc. As the beam strikes the pits and flats on the disc, it is reflected back as a series of pulses of light. These pulses are converted into electric signals. The signals are amplified and then converted into sounds, which you hear through a loudspeaker. Because light from a laser is used to play back CDs (they are not touched by a stylus), they last much longer than vinyl records.

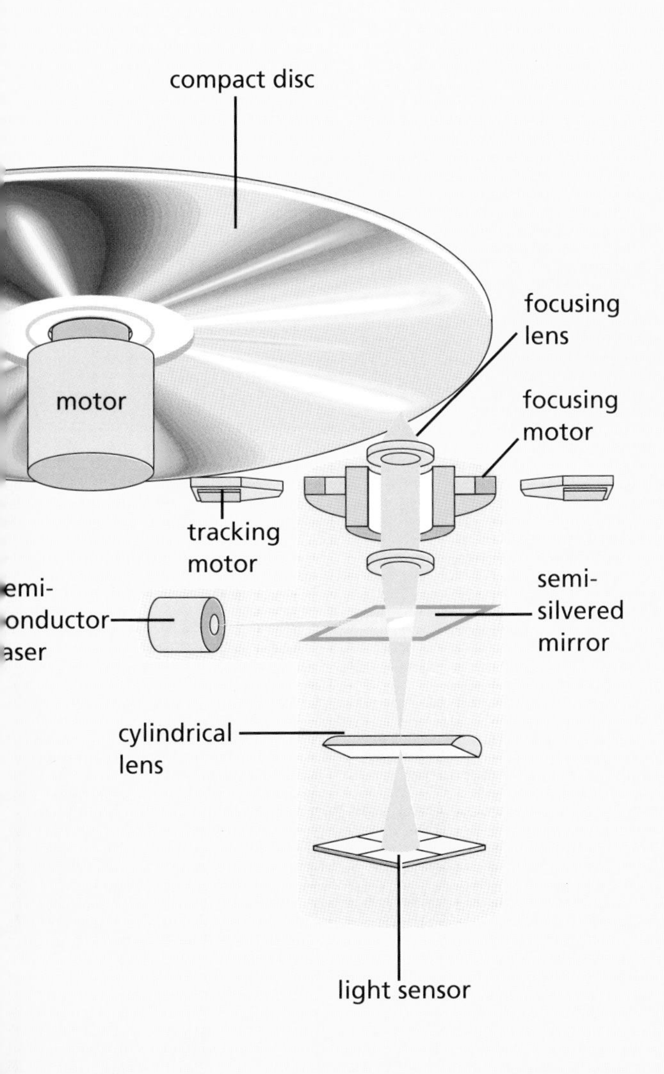

SOUNDS, WORDS, AND PICTURES

A CD stores music in a digital code. In fact, it can store any other kind of information that can be converted into a digital code. For example one CD can store more than 100 million words of text, or more than 100,000 books like this one! More and more computer programs are stored on **CD-ROMs**, and the same technique is used to store pictures. Now you can see photographs and complete movies using a computer. Multimedia, which combines pictures, sound, and words on a computer screen, would not be possible without the CD!

Two tiny electric motors in a CD player focus and move the beam of laser light that reads the disc. They make sure the laser light beam stays in the right position on the disc at all times. By using four light sensors to detect the exact position of the reflected beam of light, the motors can quickly refocus and realign the laser.

AT A MUSIC CONCERT

Watching a live music concert among an audience of tens of thousands of people is much more thrilling than just listening to a recording. Mountains of loudspeakers produce sounds that are loud enough for everyone to hear. The engineers arrange dazzling lighting effects with lasers, and the performers put on a show to remember.

Sound and light engineers have to coordinate every second of a musical concert.

Catching Every Sound

At a concert, sound engineers carefully position many microphones to pick up the sounds made by the various instruments and performers. When sound waves reach a microphone they cause tiny mechanical movements, which the microphone converts into a pattern of electric signals. These signals are very weak, and they must be made much stronger before they can travel to the loudspeakers.

The pattern of electric signals produced by a microphone go hand and hand with the sounds it receives. This pattern is made much stronger by an amplifier. Radios, TVs, and CD players all have amplifiers inside them. The amplifier uses many transistors, which are like tiny electrical switches. Transistors use a weak electric current to control a big one. And this is just how a sound is amplified. The weak electric signals from the microphone control the more powerful ones produced by the amplifier.

Is It Loud Enough?

A loudspeaker is an incredible device that converts electric signals into sounds. A key part is a coil of fine wire that is attached to a cone inside the loudspeaker. A changing pattern of strong electric signals from an amplifier is fed into the coil. These signals turn the coil into a magnet. Because the coil is in the magnetic field of a magnet in the speaker, the coil moves. The changing electric signal from the amplifier causes the loudspeaker cone to move backward and forward, making sound waves. The cones of powerful loudspeakers are made from Kevlar, a plastic material that is even stronger than steel.

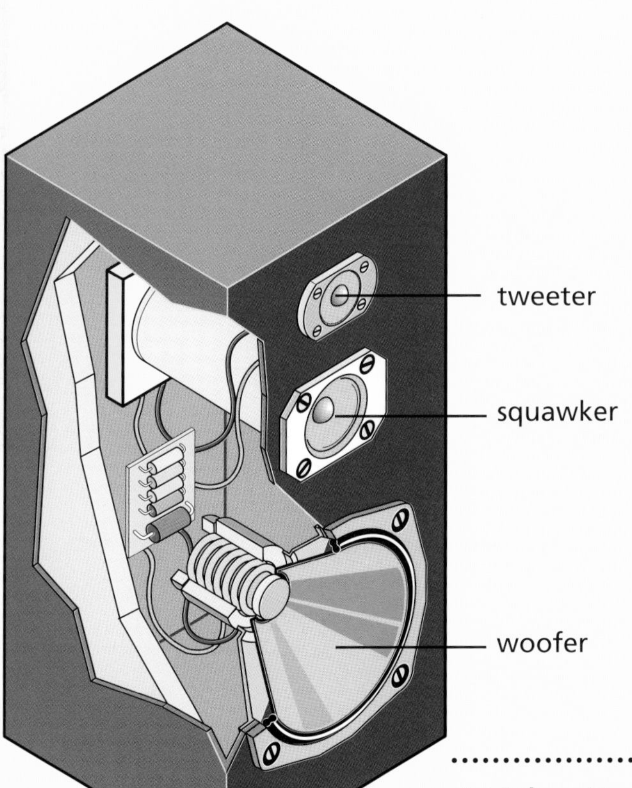

LASER LIGHT

At a laser light show bright, narrow beams from lasers move exactly in time to the music. The fantastic patterns of colored light high above the audience are controlled electronically by the sounds of the music. Because laser light can be so precisely controlled, it has many different uses in science. In industry, powerful lasers cut and drill materials. In medicine, surgical lasers are used for delicate operations such as removing damaged eye tissue. Lasers send vast amounts of information along **fiber-optic cables,** which now bring TV, telephone, and computer signals directly into our homes.

A hi-fi unit needs three different loudspeakers in the same unit. The deep, low sounds are played through a big loudspeaker called a "woofer." The middle range of sound frequencies (500 Hz to 4 kHz) are played by the mid-range unit, or "squawker." The highest notes (up to 20 kHz) are reproduced by the cone-shaped "tweeter."

GLOSSARY

acoustics relating to hearing, to sound, or to the science of sound

air pressure pressure is the ratio of force to area — a car tire, for example, is inflated to an air pressure measured in pounds per square inch. In science, air pressure is the force exerted by air on a unit surface, usually a square meter.

CD-ROM a compact disc (CD) used to store computer programs, music, pictures, or words

digital refers to a device that measures the amount of something in fixed units and jumps from one value to the next, like a digital watch

electromagnet magnet that consists of an iron core with a coil of insulated wire wrapped around it. It becomes a magnet when electric current flows in the wire.

electromagnetism the magnetic effects, for example, of an electric current flowing in a wire

electronic description of anything that depends on the movement of electrons (negative particles of electricity)

fiber-optic cable (or optical fiber) a finely spun thread of glass through which light can be transmitted. Lasers flash coded messages of light that carry vast amounts of information through the optical fibers.

frequency the rate at which something regular is repeated, for example the number of cycles that a sound wave makes every second.

friction when two surfaces rub together, friction is the force that slows their movement and produces heat

fundamental in music, the first harmonic, such as a note played on a violin or blown on a trumpet

hertz (Hz) the unit of frequency, named after Rudolf Hertz. It is used to measure the number of waves of a note that occur each second.

laser produces a highly controllable, intense beam of light of one color

microphone a device for converting sound waves into electrical signals. It is usually connected to an amplifier which, in turn, is connected to a loudspeaker, a tape recorder, or a radio.

molecule smallest amount of any substance needed in a chemical reaction

monaural describes musical recordings that were made before stereophonic recording was possible. There was only one sound track on the recording.

pitch the property of a note that makes you think it is a high or a low note or somewhere in between

resonance forced vibration of an object by a regular driving force, such as vibrations from another object

speed of sound the speed at which sound travels through a gas, liquid, or solid. It is slowest in gases and fastest in solids.

stereophonic describes a musical recording made with two or more sound tracks. The sound has depth, and you can tell where each sound is coming from.

tension the tightness of the string of a stringed instrument or the drumhead of a drum. The tension is adjusted in order to tune the instrument.

vibrato deliberate repeated changing of the pitch of a note. This helps to stress the note.

vinyl a plastic material from which the first long-playing records were made. It is supposedly unbreakable.

FACT FILE

- The overture "1812," composed by Tchaikovsky, celebrates the retreat of Napoleon's army from Moscow. Near the end of the piece, some orchestras fire real cannons in time to the music. Most orchestras make do with drums and cymbals.

- Some church bells are rung according to mathematical rules. These make sure that the order in which the bells are rung is always different.

- The chanting by the workers in Egypt who operate the shaduf—the bucket on a stick that lifts water from the Nile river—is thought to be the oldest song in the world.

- The metronome was invented by Johann Maelzel, a friend of Ludwig van Beethoven, in 1816. Its ticking sound creates an exact tempo (time) for a musician to follow.

- Akio Morita, the president of Sony, is a keen golfer and music lover. He wanted a lightweight, compact device so he could pursue both hobbies at the same time. So in 1979 the personal stereo was invented. Millions have been sold since.

FURTHER READINGS

Blackwood, Alan. *The Orchestra.* Millbrook Press

Kaner, Etta. *Sound Science.* Addison-Wesley, 1991.

Paker, Josephine. *Music From Strings.* Millbrook Press, 1992.

Paker, Josephine. *Beating the Drum.* Millbrook Press, 1992.

Staples, Danny, and Carole Mahoney. *Flutes, Reeds, and Trumpets.* Millbrook Press, 1992.

INDEX